SMOKING ALLOWED

A Pictorial Past of Honey Bee Smokers in the United States

© Paul Jackson

NB

Northern Bee Books

ABOUT THE AUTHOR

Paul Jackson retired from the Texas A & M University in 2013. He was with the University System for 44 years of which 36 years were as the Chief Apiary Inspector for the state of Texas. During this period of time, his agency regulated the movement of honey bees and/or equipment, packaged bees, queens and other bee items throughout the United States and foreign countries. Since 1976, he has been collecting antique, foreign and regular smokers. His collection is viewed by many beekeepers as one of the best in the United States.

He began collecting antique smokers due to his concern for maintaining the heritage of beekeeping in this country. He has written this book because of his desire to share his collection with those interested in the history, and the romance, of the beekeeping industry.

SMOKING ALLOWED

A Pictorial Past of Honey Bee Smokers in the United States

© Paul Jackson

ISBN 978-1-908904-46-1

Cover image by Jonathan Burbidge

Published by Northern Bee Books, 2014
Scout Bottom Farm
Mytholmroyd
Hebden Bridge HX7 5JS (UK)

Design and artwork: D&P Design and Print
Printed by Lightning Source, UK

CONTENTS

FROM THE PUBLISHER

Smoking Allowed is the first book of its kind. Paul Jackson has spent years gathering these antiques and the information he offers about each model.

This collection is by no means complete, however, though it is one of the best that exists. But even the best collections, whether in the hands of private individuals or public museums, lack some aspects of U.S. beekeeping history. We have already lost much of our heritage.

The few collections that exist have managed to save some of the best and brightest ideas, innovations and everyday tools used by beekeepers since antiquity. However, none of these are available to more than those who are close and know of their existence. And, without concerted efforts to preserve what we still have, such as this book, even that which still exists may be lost - forever.

THE GOLDEN YEARS

Reverend L.L. Langstroth's discovery of the bee space in 1851 marked a definite turning point in the history of beekeeping. This alone probably contributed more to the beekeeping industry than any other single event. With the introduction of the Langstroth hive, there was a rush to invent new and improved patterns because most beekeepers still failed to understand that the fundamental basis of the hive was the inclusion of bee space between the parts. Thus was born the market for the modern hive.

Prior to the invention of foundation, beekeepers had great difficulty obtaining straight combs and controlling the building of drone cells. To get a hive filled with good, straight combs required close attention. It was a common practice to place an empty frame between two well-built combs and let the bees build a new one between these, utilizing the bee space phenomenon. The invention of foundation is generally credited to a German, Johannes Mehring, who first succeeded in producing a crude product in 1857. He invented pressed wax wafers with the indentations common to the bottoms of the cells.

It was in 1865 that Major Hruschka, from Vienna, gave his son a piece of comb honey on a plate. Placing the honey in a basket, the boy swung the basket around his head. When Hruschka saw that the honey was thrown from the comb by the motion, the idea of the extractor was born. Once American beekeepers learned of the principle involved in the extractor (centrifugal motion) there were many attempts to make others. There were no manufactured machines available, so there were as many models as there were beekeepers to build them.

The changes were slow in coming - the invention of the modern bee hive, wax foundation and the extractor - and even beekeepers with the advantages of these implements still had difficulty controlling bees. It was known that directed smoke would calm honey bees, but only crude methods of smoking bees were in practice. There was no way to produce enough smoke to subdue an irritated colony, and beekeepers were compelled to endure a great deal of physical punishment.

Various kinds of pans or other devices were used to hold small fires made from material which smoked freely. The smoke was moved across the hive by means of blowing or waving the hand. Beekeepers often made themselves dizzy attempting to control bees in this manner.

Prior to 1875, no practical method of using smoke efficiently had been found. Tobacco smoke blown from the mouth was about the best thing available at the time. Moses Quinby is generally credited with the invention of the first practical bee smoker. This bellows-driven smoker was a great step over previous methods of applying smoke to the bees.

Now, with the invention of the bellows-driven bee smoker, the Golden Age Of Beekeeping drew to a close.

WOODEN STICK

Primitive Beekeeping

Centuries ago, man discovered that stings were less likely to occur when smoke was applied to a colony of bees. Although the reason wasn't understood, honey robbers realized that smoke forced the guard bees inside the colony and that the temperament of the colony became more docile. It was also learned that when the weather was cool, the bees tended to cluster together, making work difficult, if not impossible, and robbing impracticable. Beekeeping is an ancient practice, fascinating man from the beginning of time. Initially, honey and wax were robbed from wild bees and used for sweetening foods, making medicines and providing wax for candles. Later, when bees were kept in pottery vessels, wooden boxes, straw skeps and other primitive containers, the major beekeeping tool was a smoking stick from an open fire.

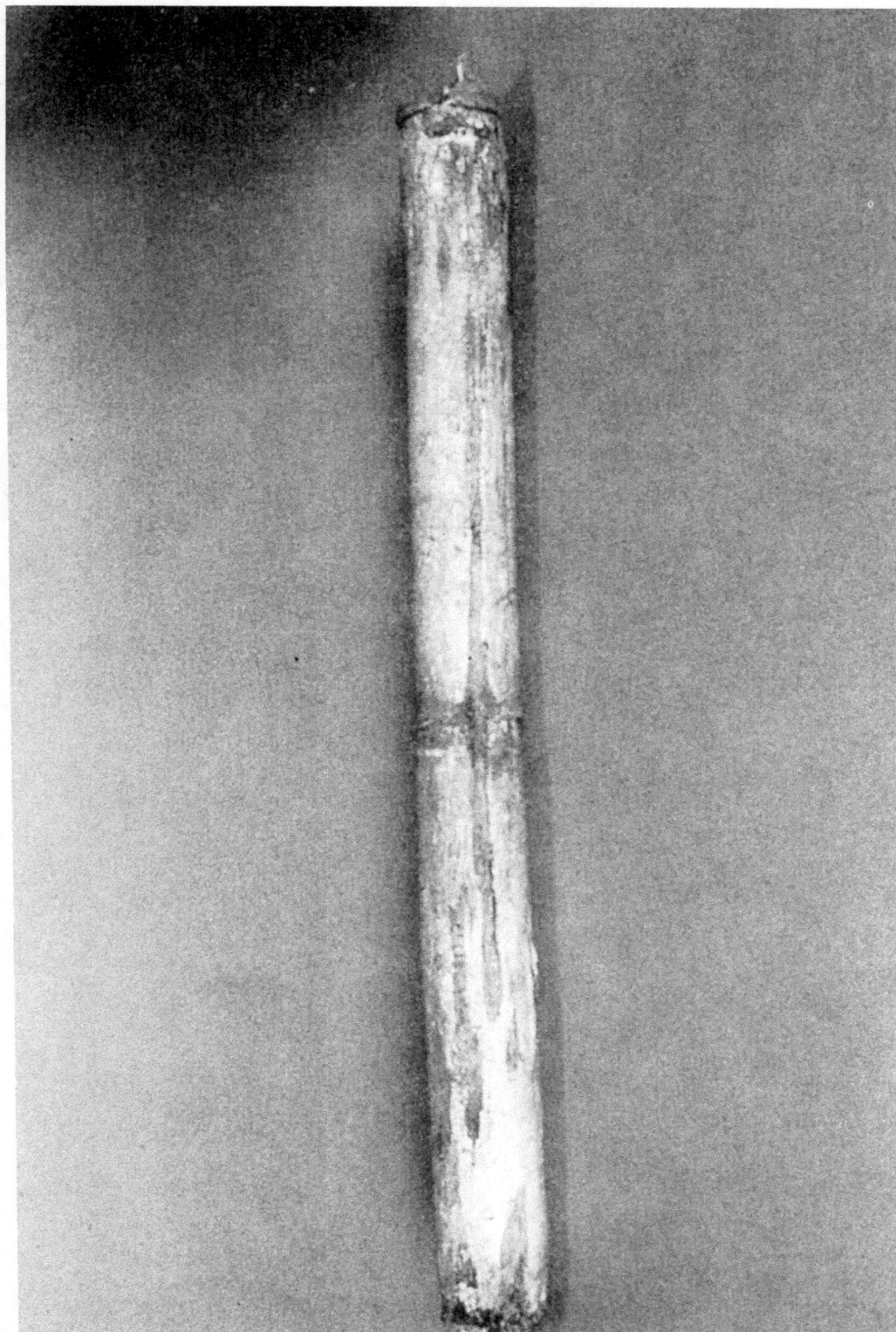

SMUDGE POT

Mid-1860s

This particular smoker was common among beekeepers prior to the addition of the bellows. The construction was generally that of a dome-shaped cap with vent holes on top. Usually, a large pot or fire container was inside, having air holes on the side of the pot and on the bottom as well. This early smoker was crude in appearance, but was properly ventilated to insure a good draft. The smudge pot was placed on the windward side of the hive so smoke would blow over the frames. For really solid, subduing smoke, this was a good bellows-less smoker, but the large quantities of smoke produced were more than necessary and often unpleasant to beekeepers working nearby. However, there was a tendency for these primitive smokers to burn out quickly and to occasionally blow sparks among the bees. The wire hook was used to hold the smoker on the side of the hive.

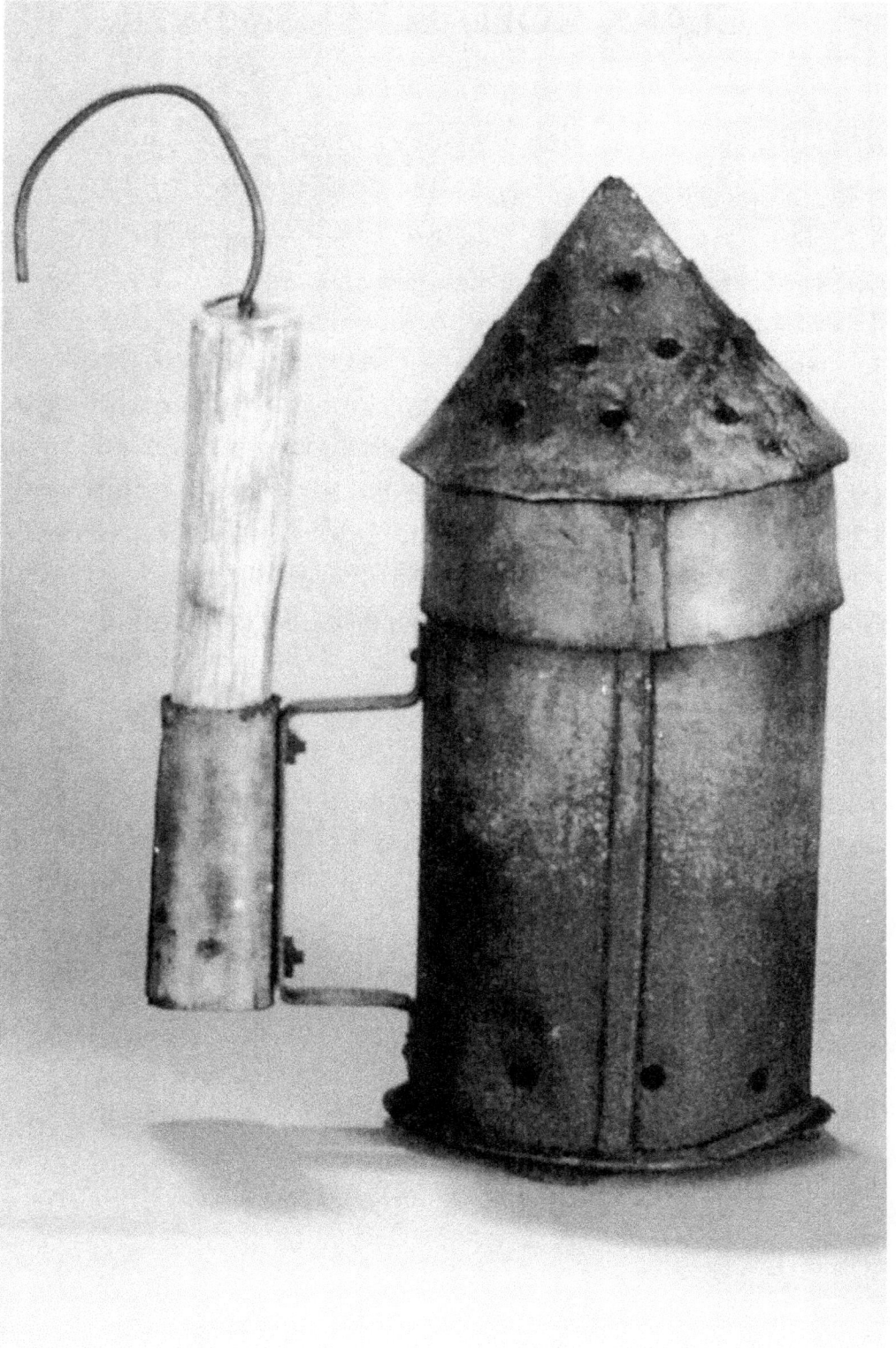

CLARK COLD-BLAST SMOKER

1879

The "cold-blast" principle was invented almost simultaneously in 1879 by J.G. Corey of Santa Paula, California, and by Norman Clark of Sterling, Illinois, each without knowledge of the other. Clark's was the better design, and further improvements led to the Clark Cold-Blast Smoker. The "cold-blast" principle meant that the air was forced directly from the bellows by means of an internal tube to a point inside the fire box, but above the fire. This made it possible to send cold, smoke-charged air out the spout onto the bees. The disadvantage of this smoker was that when set down for a short period of time, it would readily go out. This meant the smoker had to be relit for nearly every hive. This smoker was so popular, however, that over 20,000 were sold in a short period of time. The Clark Smoker was sold for less that half the price of others on the market, and it could send a stream of cold smoke over eight feet. Many beekeepers preferred this smoker because a single filling lasted longer and gave a heavy, cool smoke. Also, it could be replenished from the bottom, leading to its also being called a "breech-loading" smoker.

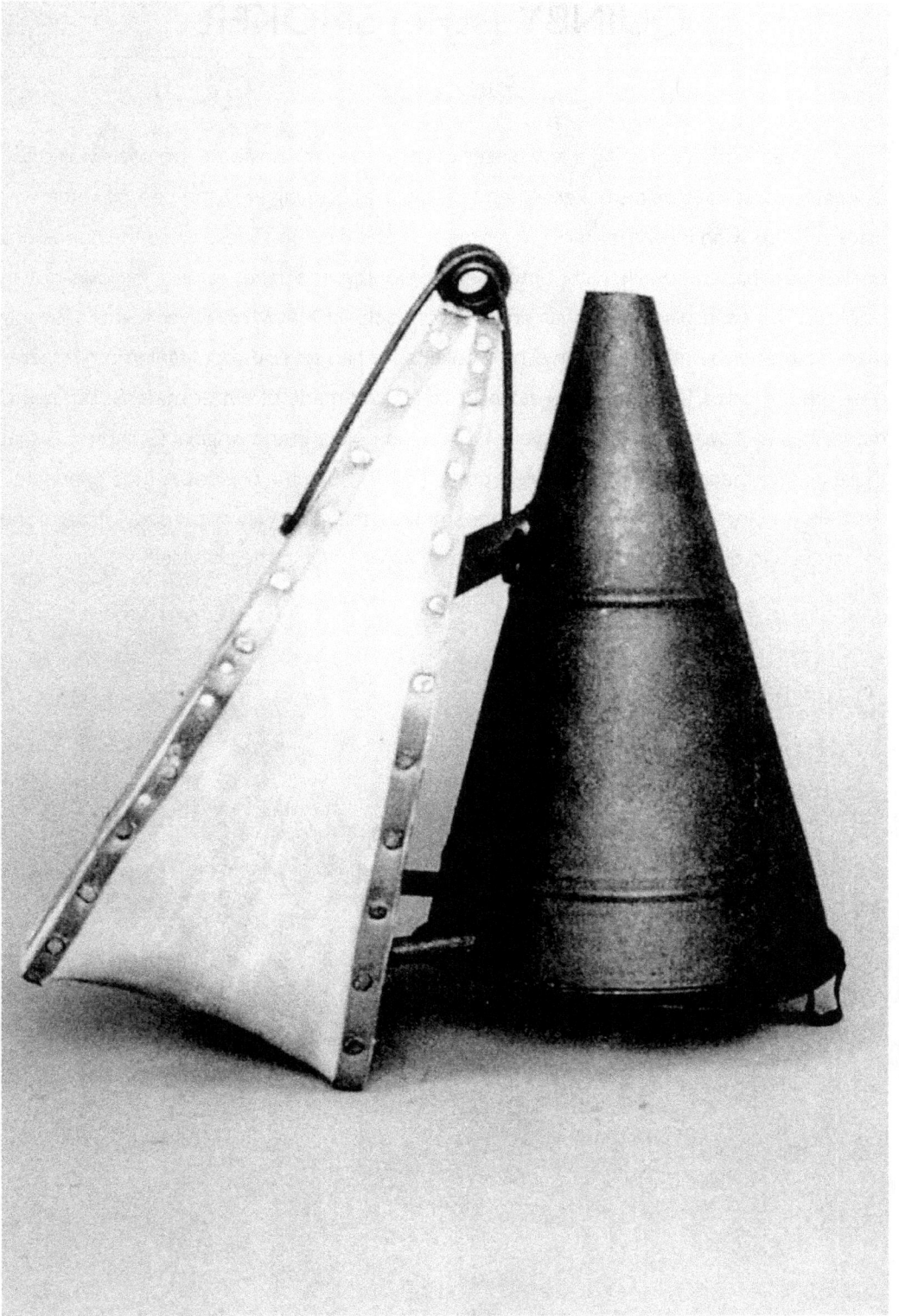

QUINBY (style) SMOKER

Early 1800s

A pan or tea kettle containing a few pieces of rotten wood and some live coals was often used as a substitute for a smoker. When this fuel was burning well, the pan or kettle was placed so the wind blew the smoke over the top of the colony. These were the conditions beekeepers had to work with until the development of the earliest bellows-driven smokers. This particular model was similar to the actual Quinby. However, it was probably manufactured by another company. This smoker style had several important improvements over other models. It was not only stronger, larger and made of better material, but it was made so that it would not go out very easily. One defect in the original Quinby was that there was no means of allowing a direct draft of air through the fire pot. A tube connected the bellows directly to the fire pot. In later smokers, this tube was dispensed with, and the air was blown directly into the fire pot through an opening in the pot itself.

THE QUINBY SMOKER

1880s

The Quinby Smoker appeared on the market in the 1880s. Although Moses Quinby never patented this smoker, one was obtained on March 11, 1879 (patent no. 213,251), by his son-in-law, Lyman C. Root of Herkimer County, New York. Presumably, because the bee smoker was a novel item, the patent was classified under an "apparatus for destroying insects by fumigation". The Quinby Smoker came in different sizes and designs. The model pictured here is 11 inches high with a 2-1/2 inch diameter. The brackets attaching the fire chamber to the bellows are made from castiron. The bottom of the smoker is castiron and also forms part of the lower bracket. The use of castiron parts in smokers is rare.

BINGHAM (style) SMOKER

about 1880

This particular model looks very similar to an actual Bingham Smoker. However, it was probably manufactured by another company. The fire cup is rather large in order to hold enough fuel to burn longer than similar-time smokers, since earlier fire chambers usually were comparatively small. The top was funnel-shaped, which meant the smoker had to be turned upside down to direct smoke onto the bees. This occasionally caused fire to drop or even shoot onto the bees while smoking the hive. This was termed the "hot-blast" system, since the blast of air from the bellows was blown directly through the fire from below. This made a large volume of smoke, but sometimes beekeepers worked the bellows so vigorously that it actually blew fire from the nozzle. This was another reason for calling this the "hot-blast" system. This type of smoker had some other defects, too. It was top-heavy, so it had a tendency to topple over and often flopped open at the most inconvenient times. Another defect was that the bellows was pear shaped and hinged at the narrow end. After several months of use, the hinge became loose and the bellows less efficient in forcing air into the fire chamber.

BINGHAM SMOKER

about 1880

The earlier Bingham smokers had two basic characteristics that separated them from other smokers - a wire coil on top of the spout for directing smoke and a wooden stick or strip that ran the length of the bellows that the twisted-iron wire was attached to that fastened the bellows to the fire chamber. The Bingham was strong, well made, did not clog up and burned any fuel. It was improved later by adding a curved spout to prevent fire or ashes from dropping onto the bees, and a safety attachment was added to prevent burning the fingers when removing the hot cone for refilling. "Fire dropping" is a term used when sparks go out the top of a smoker. Many early smoker designs had to be turned upside down in order to apply smoke to the hive. The Bingham Smoker came in several different sizes. The larger sizes had wide shields, while the smaller sizes had narrow shields. All had wire handles or wire coils to hold the curved spout in place. This was used to direct smoke onto the hive.

CLARK COLD-BLAST SMOKER

1882

When it was necessary to fill Quinby or Bingham smokers, the top, or spout, had to be removed. There was a good chance that you'd burn your fingers removing the top or putting it back on again. The Clark Smoker, however, was made so that it was filled from the bottom of the fire chamber. The door, on the bottom, was riveted on one side, so when it was necessary to open for refilling, it simply slid around with a light touch of the fingers. An additional advantage was that if a bit of additional draft of air was needed to make the fuel burn more rapidly to give an increased amount of smoke, the door could be turned, or opened, to leave a small vent. This particular Clark smoker has a single ridge around the barrel of the fire chamber. One ridge separates this model from earlier versions, since they had two ridges running around the chamber. The instructions that came with this model indicated that several small holes could be drilled just below this ridge to allow additional air to be drawn in for the fire chamber.

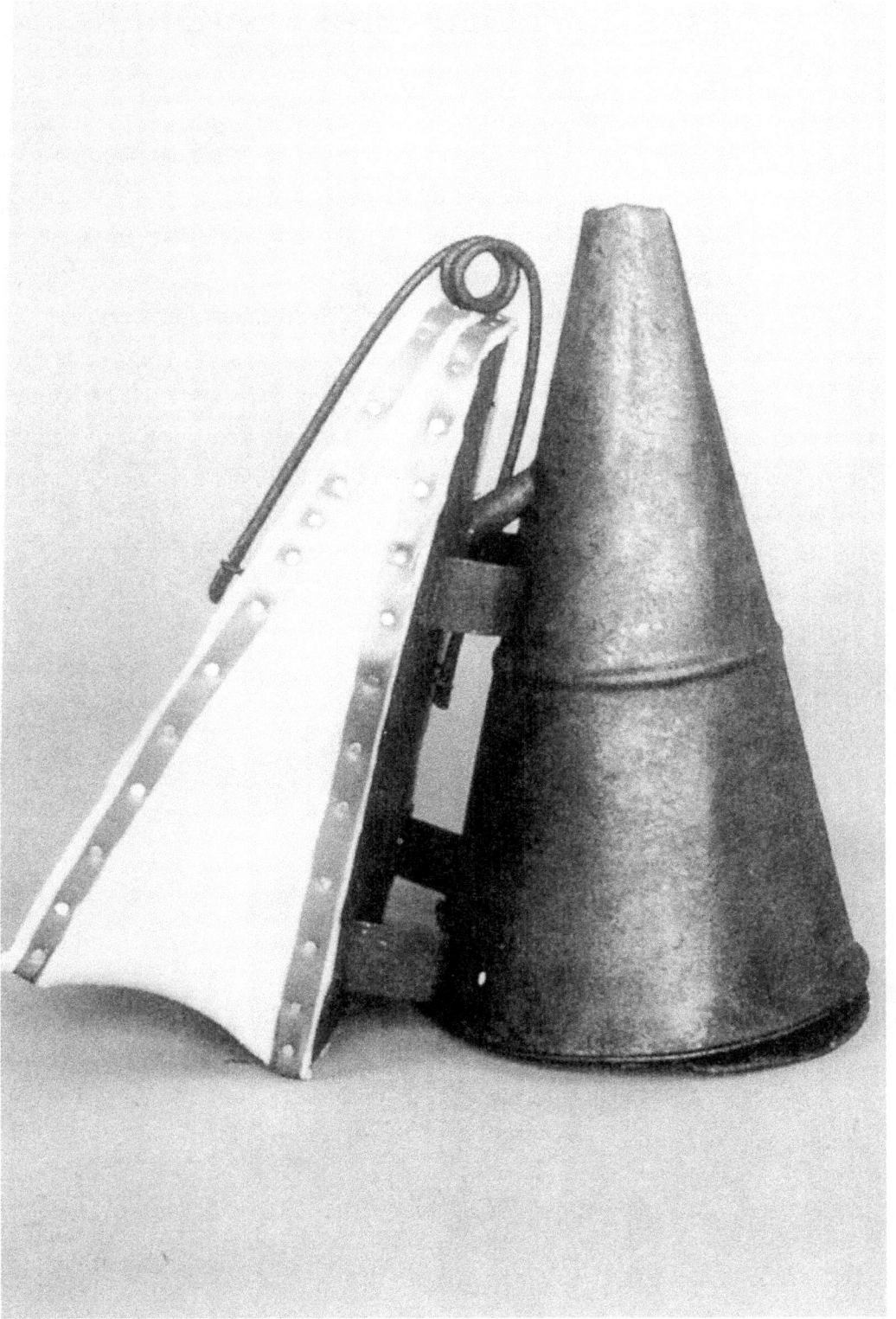

SIMPLICITY COLD-BLAST SMOKER

1883

This smoker's shape is unforgettable. The large "base" made it virtually impossible for this smoker to overturn. It was made so that there was little danger of any fire or ash escaping while the smoker was being worked. With the Clark Cold-Blast Smoker there had been trouble with burned fingers when removing the top to refill the fuel chamber. The Simplicity Smoker had a sliding door on the side of the chamber used for refilling. This slide also opened and closed the damper, which was simply a ring of tin surrounding the first receptacle with a wooden handle to move it. The damper opening could be adjusted in order to draw more, or less, outside air, preventing the fire from going out when the smoker was set down for short, or even long, periods of time. Going out was the major complaint about the Clark Cold-Blast Smoker.

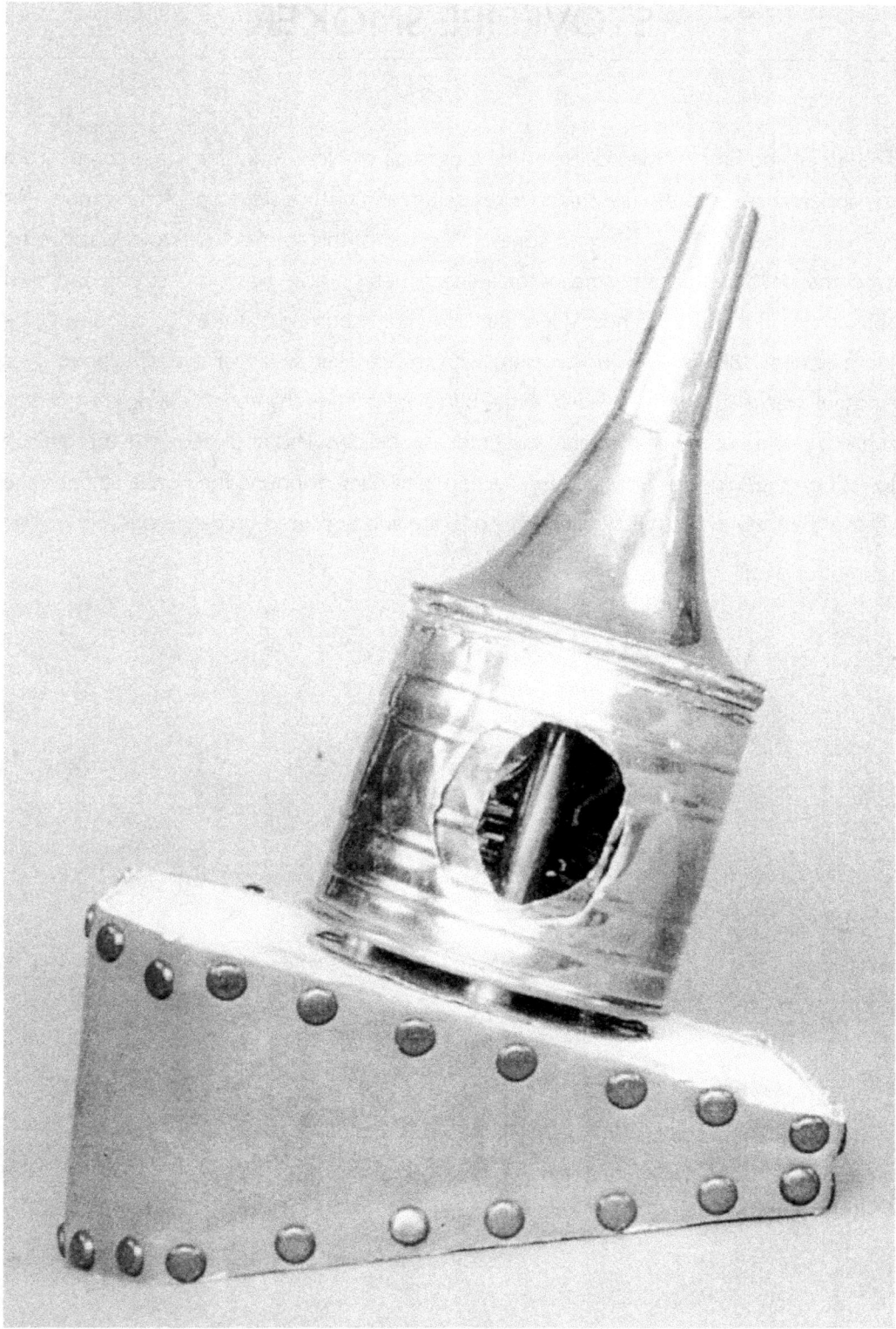

STOVEPIPE SMOKER

1886

The Stovepipe Smoker was designed by Coggshall of New York. The name comes from the appearance, since the fire chamber resembles a miniature stovepipe. A flat handle was riveted on the side of an inverted cone on the top of this smoker. The stove was riveted onto the bellows board by means of a sheet-metal brace, giving it a strong and rigid appearance. But without a heat shield, the stove could burn out at the bottom, destroying the attached bellows. A major problem with this smoker was that it easily tipped over. Another was that the top fitted over the barrel of the fire chamber, allowing creosote to run down the outside of the tube and enter the bellows. This also made the top difficult to refit on the "pipe" and it was often necessary to bang or pound the nozzle to open the smoker. Basically, it was messy, difficult to operate and bordered on dangerous.

PUMP SMOKER

about 1890

This is a bellows-less smoker that, when operated as intended, produced a solid, subdued stream of smoke. However, such quantities of smoke were usually unnecessary and often unpleasant to the beekeeper. The outer sleeve had to be hand pumped, back and forth, to create the draft for this smoker. This action was definitely distracting for several reasons. First, it required a lot of rapid hand motion, and at the same time it made a lot of noise. These combined actions could, and often did, result in many stings to the operator. Nor did this smoker have any sort of anti-sparking blast-tube device to prevent sparks from flying out, which could set fire to any nearby clothing or to the colony itself. Since it took two hands to operate this smoker a second person was often required when working bees. This activity - running the smoker - often fell to the wife or son of the beekeeper. And, since the person operating the smoker was many times the target of colony defense (while the beekeeper enjoyed the protection of the smoke), this smoker did not last very long.

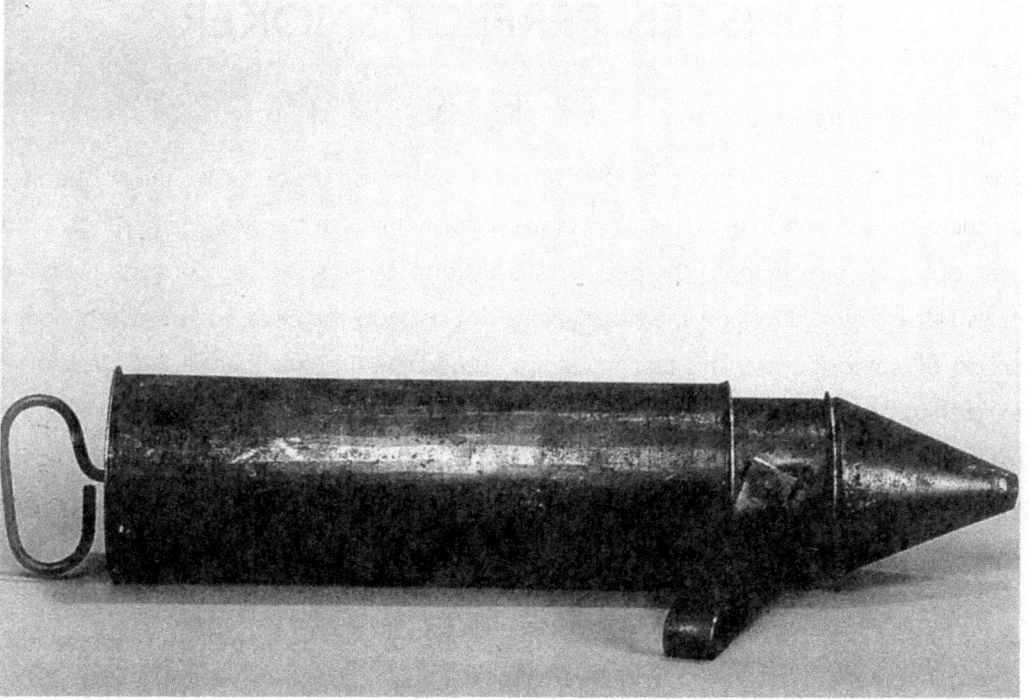

FUNSTEN PERFECT SMOKER

about 1890

This smoker was manufactured by the Funsten brothers of St. Louis, Missouri. It operates on the hot-blast principle, that is, air is blown directly through the fire, which sends a hot blast of smoke directly onto the bees. This is liable to cause sparks, cinders and even live coals to be blown down on the hive, landing not only on the bees, but on comb, honey, brood, eggs and the like. This technique also heated the fire chamber to extremely high temperatures, rendering it difficult to work around and shortening its life. This was the major disadvantage of the Funsten Perfect Smoker. Interestingly, it was also used by the fur industry for driving animals from their dens. A strip of burlap about eight or nine inches wide and about 30 inches long was sprinkled with sulfur powder, rolled up tightly and placed in the smoker. After burning for a minute or two, the bellows was attached and worked until a fire was going well. Then the small end of the fire chamber was placed into the den, and two or three blasts of sulfur-laden smoke were directed inside. The animal subjected to this treatment would escape from the fumes, leave the den and could be captured. This practice would probably be frowned upon by some animal rights groups.

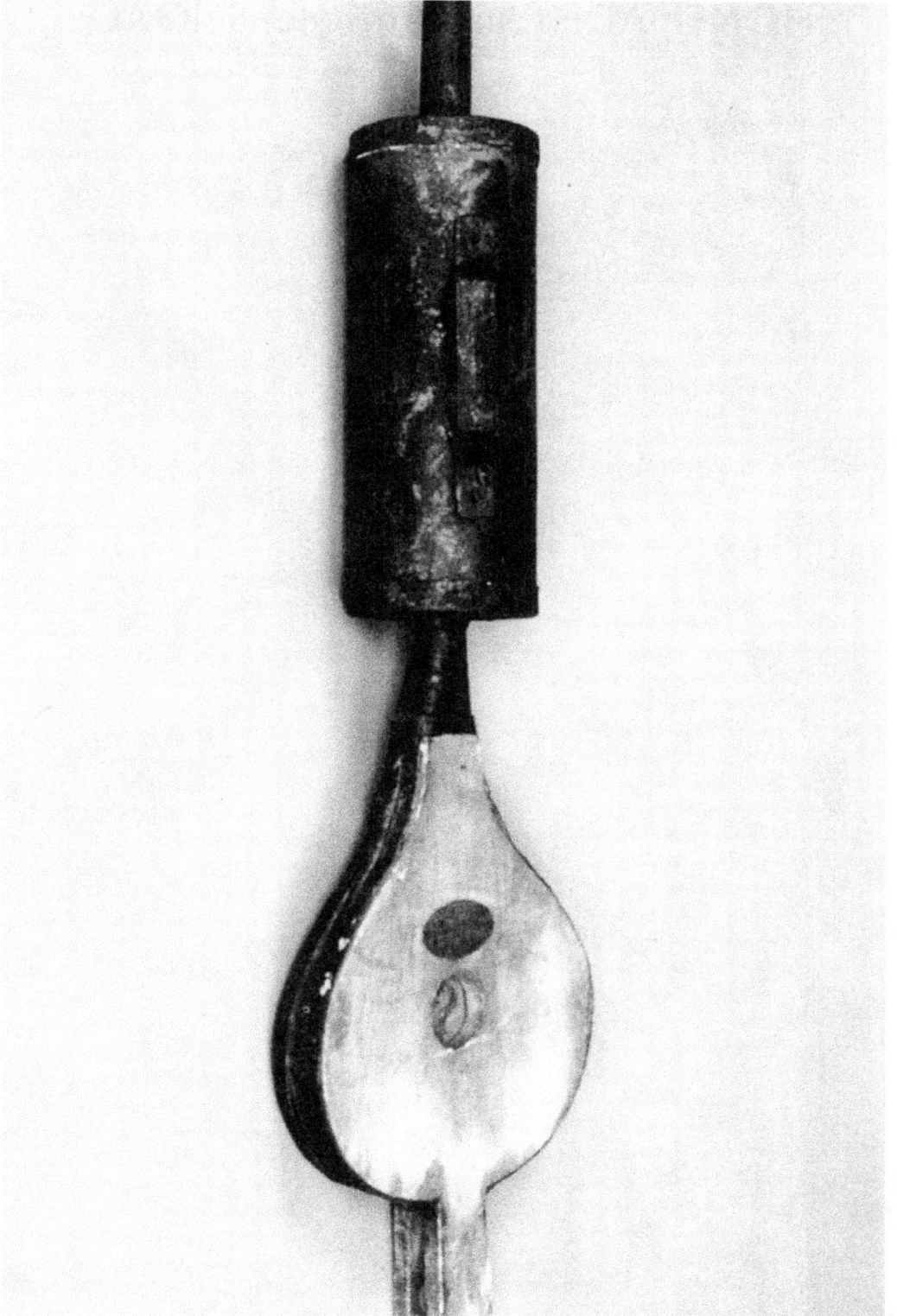

HOMEMADE (Bingham-style) SMOKER

about 1890

Before smokers were commonly available, beekeepers often made their own, modeled after drawings in catalogs or examples from friends. This is an example. For good, often too-good ventilation, holes were made in the bottom and often in the top. All those extra air holes meant that these primitive smokers worked on the hot-blast principle that is, the draft of air was blown directly through the fire. This often led to over-heating the fire chamber. Also, because they worked on the hot-blast technique, and they were overworked, burning colonies and beekeepers' clothing were too-common occurrences.

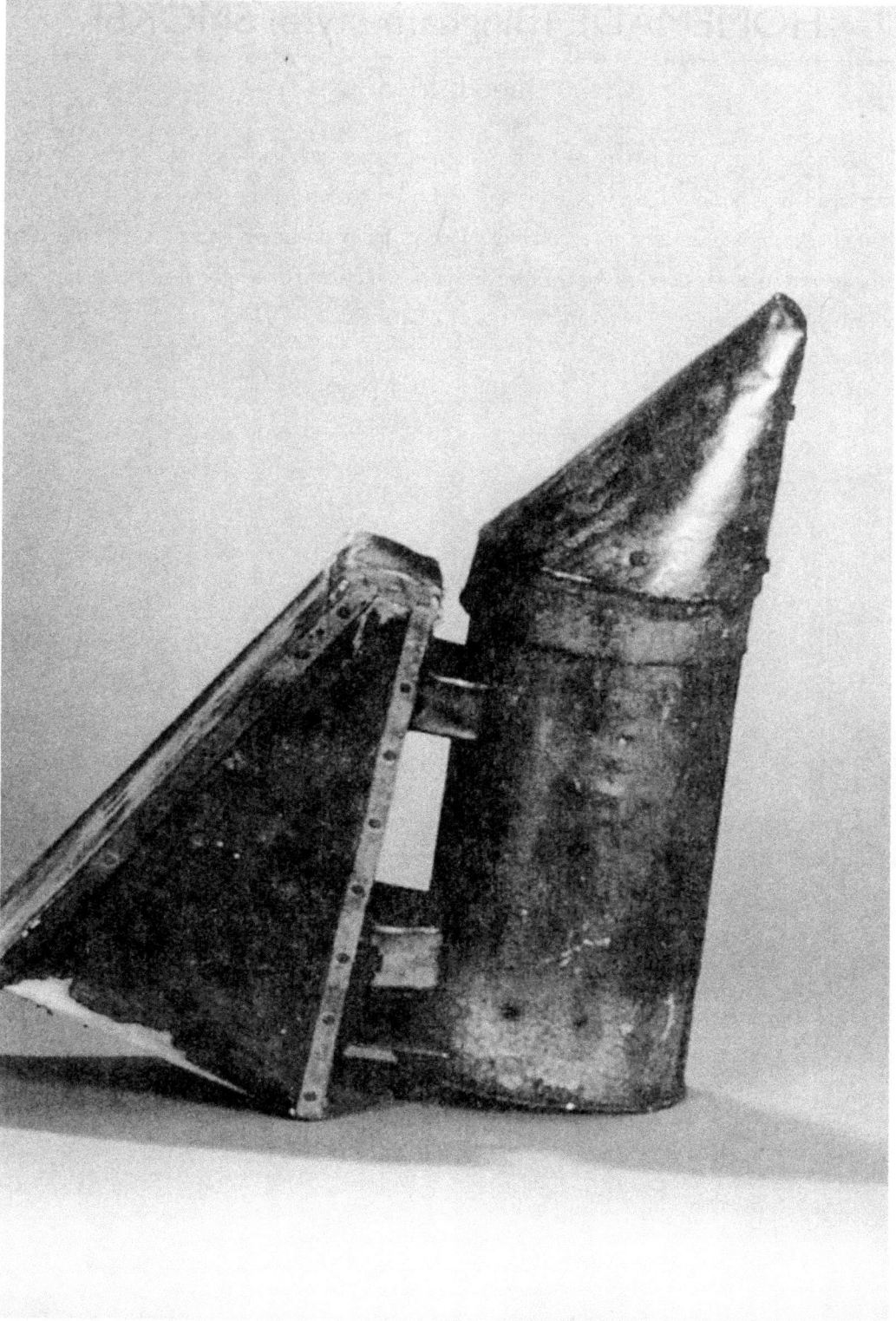

HOMEMADE (Bingham-style) SMOKER

about 1915

This smoker was probably made from a number-two peach can. The top of the can was removed, and another can was attached and used to funnel the smoke up and out. A handmade wire was used to help prevent coals from dropping and to aid in directing smoke where it was needed. Also, a wooden stick ran up and down the face of the bellows, which gave this smoker the characteristics of a Bingham.

CLARK COLD-BLAST SMOKER IMPROVED

EARLY 1890s

The Improved Clark Cold-Blast Smoker appeared to be well built and handsome. The bellows was large and substantial, with a spiral spring inside that was both elastic and unbreakable. The outside edges of the bellows were metal bound, which helped increase its life and protected the leather edges from wear. Also, the binding prevented the bellows boards from warping and held the boards firmly while the bellows was operated. This new design reduced the tension on the internal spring, helping it respond instantly with a strong blast of air when compressed. The bellows was reversed on the improved model which gave more room for handling. For best results, the fire cup should be under the bellows, the hand grasping the large end of the bellows on one side. Also, this smoker was improved by putting a set of perforations on the end of the fire cup as well as under it. But even with all these improvements, it still had the same basic problems as the old design. When setting it down for a short period of time, it would readily go out.

BINGHAM SMOKER

1892

This smoker was manufactured by T.F. Bingham of Farwell, Michigan. The fire cup was larger than those of previous smokers and was ventilated in such a way that a continuous draft could be maintained even when the smoker was not in use. Several sizes were made in the late 1800s. The fire cups on all models had an anti-spark tube situated just below the grate inside, constructed to prevent the suction of sparks into the bellows or out into the air, setting fire to clothing during bellows operation. A wire coil was on top of the smoker in order to hold the curved spout in place, which directed smoke onto the hive while it was in an upright position. This wire coil was also used when removing the top to reduce the chance of burning the fingers when opening or closing. The disadvantage of this smoker, however, was that it had an old-style nozzle, which was top heavy and had a tendency to tip over or fly open at the most inconvenient times. The copper hinge, however, was lightweight and riveted securely in place on the side of the fire cup.

BINGHAM SMOKER

1900

This smoker is very similar to other Bingham models, but the main difference is the bellows, which is well-finished, thus increasing its life by preventing warping of the boards. The thumb groove in the bellows board gave an excellent grip. The lumber generally preferred for these was basswood because it was light and not apt to split. The leather part of the bellows was smooth and soft, usually made of sheepskin or deerskin, and was fastened to the wood with glue to make an airtight connection. A quarter inch-wide metal binding was then tacked to the board for additional strength and a strong, secure fit. It also had an inside spiral spring. This was a good, serviceable smoker for the commercial beekeeper.

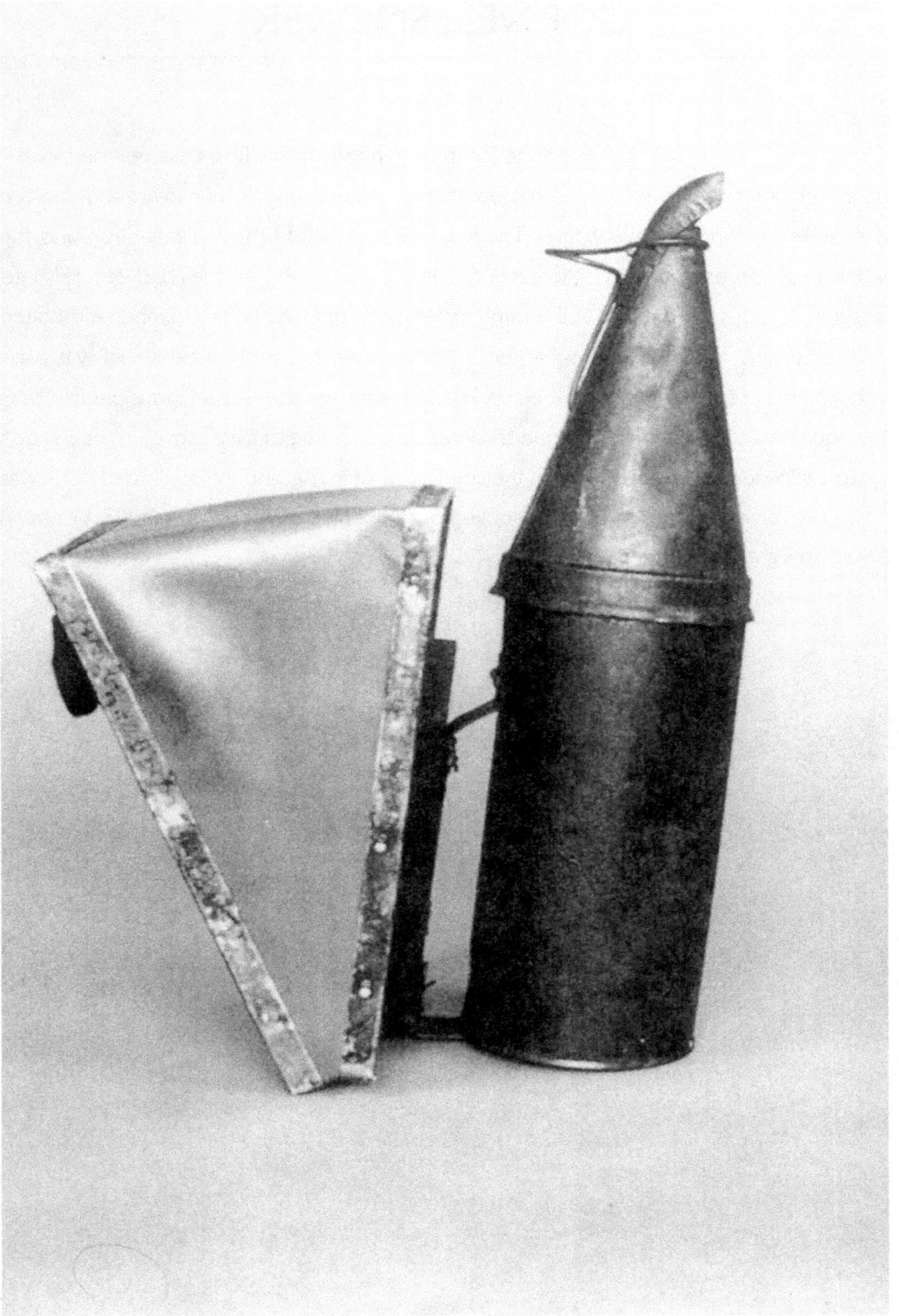

CORNEIL SMOKER

1901

The Corneil smoker was just the thing for those who wanted a first-class, reliable smoker at a medium price. The top was hinged so that it could be thrown back and the fire cup refilled without burning fingers. The hinge was made of light, malleable iron, and the working parts were milled to an exact fit so the curved spout always fell back securely into position. The legs were made of the same material, riveted to the fuel chamber and bolted to the bellows, making it impossible for anything to get loose. The heat shield was plain, with an airspace between it and the cup. The air blast used the Corneil principle, involving the addition of a supplementary tube to increase the volume of air. Grooves in the bellows boards afforded an excellent grip on the smoker while in use, and the galvanized hook was a real convenience, since it could be hooked onto the hive right side up, ready for use and always in sight.

ROOT JUNIOR SMOKER

1903

Three sizes of Root smokers were sold - the Jumbo, with a four-inch stove, the Standard, with a three-and-a-quarter inch stove, and the Junior, with a two-and-a-half-inch stove. The latter had no hinge. The Junior smoker was not only neat in appearance, but the spout held its position on top of the stove without danger of toppling over, no matter how roughly it was used. There was no reason why the spout should be large and heavy, having a capacity that was rarely, if ever, needed. This newly designed spout was compact, simple and yet served perfectly the purpose of deflecting the smoke at a working angle from the stove. A very neat wire-coil handle, which remained cool under any circumstances, was riveted securely in place on the backside of the spout at a point that was easily used for lifting or replacing it. It opened so easily that it was not necessary to bang or pound the spout remove it. The Junior was just the thing for quick inspections because it was lightweight and compact. For quick trips or inspections, many preferred the Junior.

ROOT SMOKERS

1903

Other features of the Root smokers, starting about the turn of the century, included the abandonment of the internal valve, which was liable to clog with creosote. The flexible hinge had a longer shank, permitting the top of the spout to adjust itself to the fire cup when closed. This also shoved any built-up creosote out of the way, making this a self-cleaning model. The curved spout directed the smoke at an angle most convenient for use, and the wire handle, that was always cool, made removing the top easy for refilling. The anti-spark blast tube at the bottom that accepted the air blast when the bellows was compressed increased the volume of air and prevented sparks from flying out at the rear when the bellows was released, setting fire to clothing. This feature was especially prized by women beekeepers.

ROOT COPPER SMOKERS

1903

Root smokers were generally constructed of either tin or copper. The copper smokers did not rust, but the metal was soft, and they had to be handled carefully to prevent being battered and bent out of shape. A hook was attached to the back of the bellows or the front of the smoker so it could be hung on the side of a hive, within easy reach. Copper smokers were made in Jumbo with a four-inch stove, and Standard, with a three-and-a-quarter-inch stove.

THE (WOODMAN) BINGHAM SMOKER

1903

This direct-draft smoker was manufactured by the Woodman Company under the Bingham patent. It's easy to see that it does not match the basic features of the earlier Bingham smokers, which had a wire coil on top and a wooden stick or strip running the length of the face of the bellows. The earlier Bingham smokers were generally purchased by long-time beekeepers who were very loyal to Mr. Bingham, but he knew that modern smokers were the way of the future. The Woodman model had a top that was small and compact and could direct smoke on the hive while in an upright position. The bellows was metal-bound, which increased the life and prevented the boards from warping. It was suggested that Mr. Bingham was trying to corner the market on smokers. In order to do this, he had the Woodman Company manufacture the new style smoker, which appealed to younger beekeepers, while at the same time, the original design appealed to older beekeepers. He had the patent on both and was selling to both generations of beekeepers.

KRETCHMER SMOKER I

1908

The curved nozzle of the Kretchmer Smoker was raised enough to prevent the top from burning through, yet low enough so it wouldn't be top-heavy. They were self-cleaning because the soot from the nozzle dropped back into the fire and was converted to smoke. The nozzle was fastened to the bellows with a steel butt that brought the nozzle correctly and firmly onto the fire chamber without battering or crimping the edges. The nozzle was easily opened and closed without necessarily touching it. The folded metal binding around the bellows prevented the boards from warping and held the leather tight to the wood. The connection between the fire chamber and the bellows was of stamped metal, securely riveted to the fire chamber and secured to the bellows with two bolts to brace the bellows to the nozzle. This braced the fire chamber in every direction, so that even rough use would not damage it. The large tube within the base of the fire chamber and the absence of any tube connection between the bellows and fire chamber prevented sparks from being sucked back into the bellows.

KRETCHMER SMOKER II

1908

This Kretchmer model was made in the same substantial manner as the other models, but was smaller and, consequently, cheaper. It was designed for beekeepers with 10 to 20 colonies and was referred to as the "Standard." It was strong and durable, and the connections between the fire chamber and the bellows were stamped metal. The holding brackets were riveted to the fire chamber and secured to the bellows with two bolts. The cover fit on the outside of the fire box, which made it convenient to close. However, beekeepers with 50 or more colonies and larger crops of honey would have found the Jumbo a better choice. It was only several ounces heavier than the Standard, but it held nearly twice the fuel, which lasted twice as long, gave more smoke, was easier to hold and was easier to clean. Enough said.

ROOT IMPROVED JUNIOR SMOKER

1926

The Improved Junior had a larger fire box - three and a quarter by five inches - and was made from better quality material. The bellows was made of sheepskin and was bound to the boards by a projecting metal strip. The Junior smoker had an anti-spark tube preventing sparks from being forced out of the draft tube. It also prevented smoke, sparks and fire from being sucked back into and destroying the bellows.

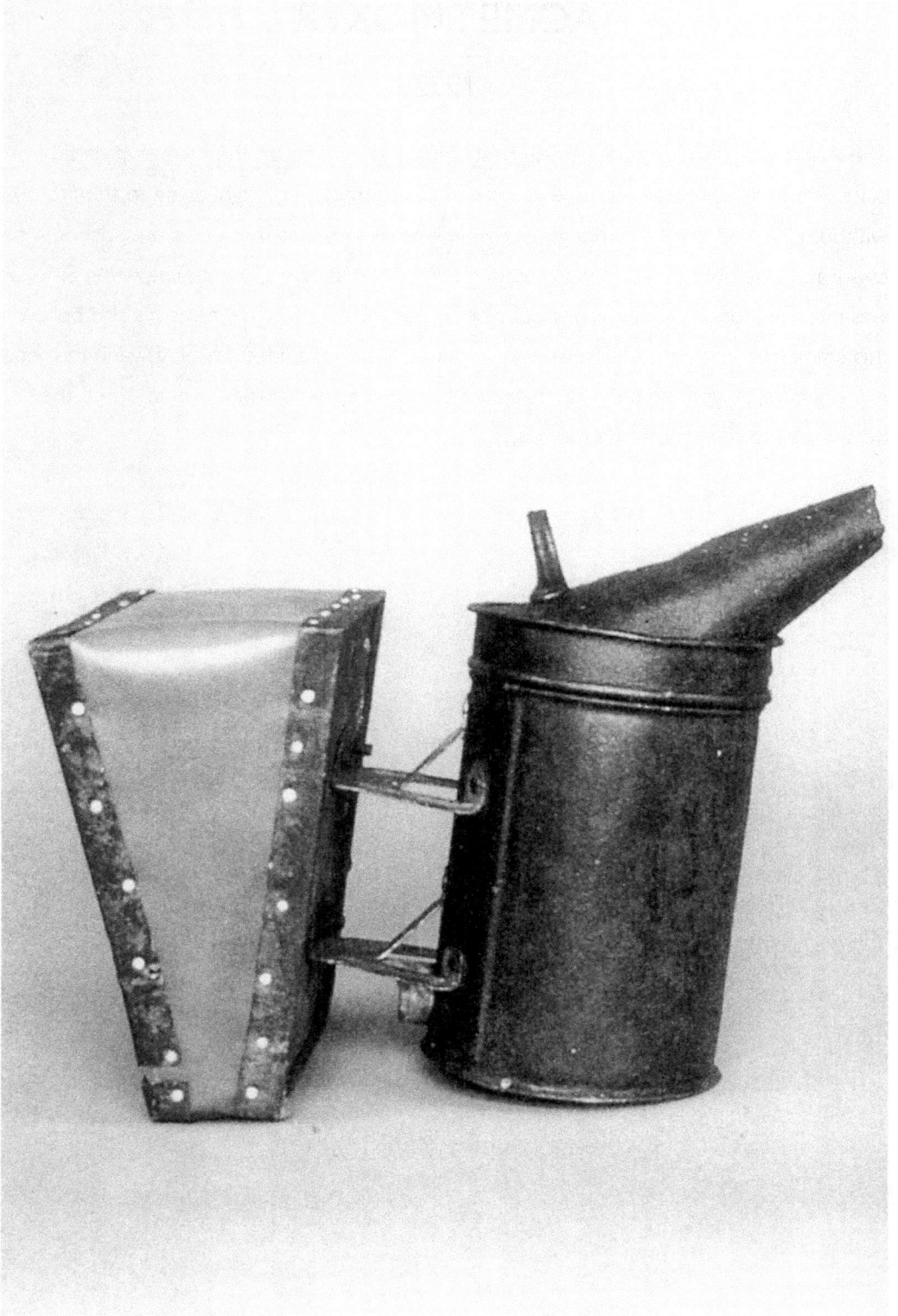

ACME SMOKER

1932

The Acme was once considered the best smoker of its size on the market. It had a hot-blast system and produced a large volume of dense smoke at a right angle to the fire box without a curved nozzle. Although its size was small two and three quarters inches wide by six and a half inches tall - it was economically priced at only 70 cents. The Acme Smoker was intended for the amateur or hobby beekeeper with only a few hives. It's also believed this smoker is a Higginsville (see next) product. A bar of folded tin or galvanized metal, running parallel with the bellows, protected the hand from coming in contact with the fire box, similar to the Higginsville models.

HIGGINSVILLE SMOKER

1932

This smoker, as its name implies, is strictly a Higginsville product. It had a hot-blast system with a three-and-a quarter-inch barrel, and the nozzle was hinged to allow filling or cleaning. It had as strong a blast as any smoker on the market. A bar of folded tin or galvanized metal ran parallel to the bellows, protecting the hand from coming in contact with the fire box. The Higginsville Smoker was guaranteed to bum any fuel, including sound, hard wood, corncobs, rotten wood, planer shavings, peat, old rags and greasy waste.

GLOSSARY

Antlspark tube

A hollow, elongated cylinder or tube or device that prevents a spark or flash from entering the bellows.

Bellows

An instrument that, by alternate expansion and contraction, draws in air through a valve or orifice and expels it through a tube into the fire chamber.

Coiled wire handle

A spiral or series of wire loops that form a coil to hold in place a curved spout on top of a smoker for directing smoke. The coil is also used to remove the top without burning the fingers.

Flexible hinge

A jointed or flexible device on which a lid, top or other swinging part is attached to the fire chamber.

Grate

A frame of parallel or crossed bars holding a bed of burning material that allows air to pass through from below and enter the fire chamber.

Hook

A curved or bent device for holding a smoker on the side of a bee hive.

Locked nut

A perforated block, usually metal, that has an internal screw thread and is used on a bolt or screw for tightening or holding something.

Stamped metal legs

A forked or jointed metal device that attaches the bellows to the stove or fire chamber, made by stamping an originally flat piece of metal.

Stove

A portable or fixed apparatus that bums fuel in order to provide heat, and thus smoke, for a special purpose (as in heating air in a hot-blast). Usually a pot or fire container sometimes called a fire chamber, fire cup, fire pot or fire box.

Top

A cylindrical or conical device that has a tapering constriction to speed up or direct a flow of smoke.

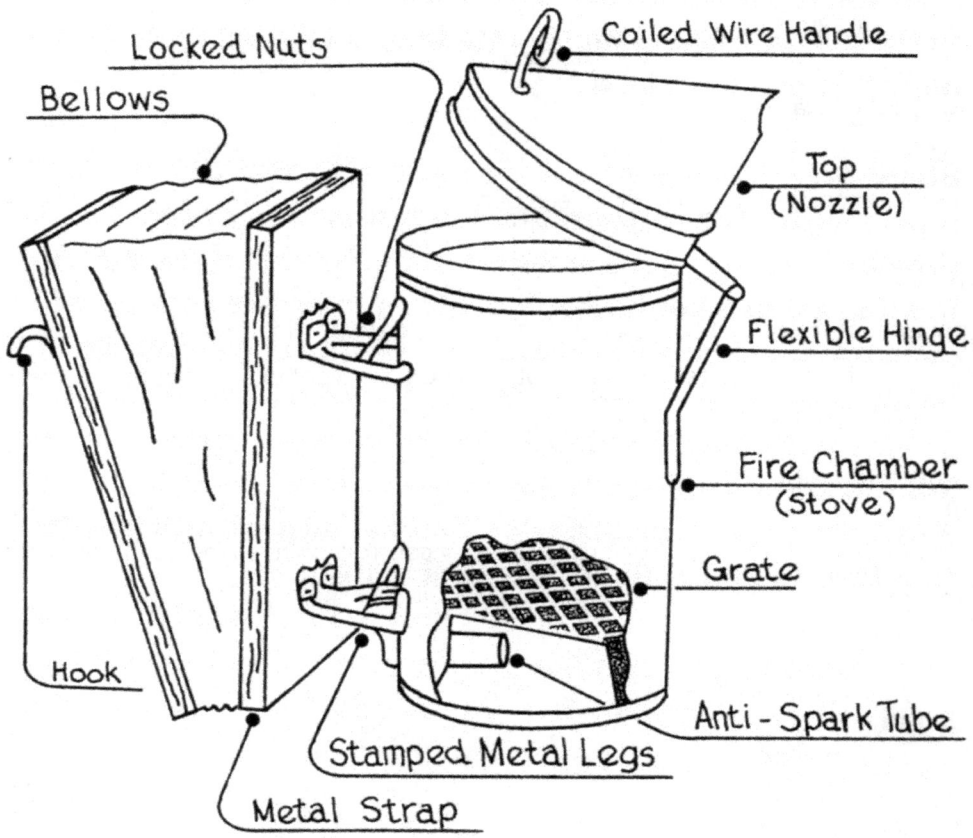

Locked Nuts

Bellows

Coiled Wire Handle

Top
(Nozzle)

Flexible Hinge

Fire Chamber
(Stove)

Grate

Hook

Anti - Spark Tube

Stamped Metal Legs

Metal Strap

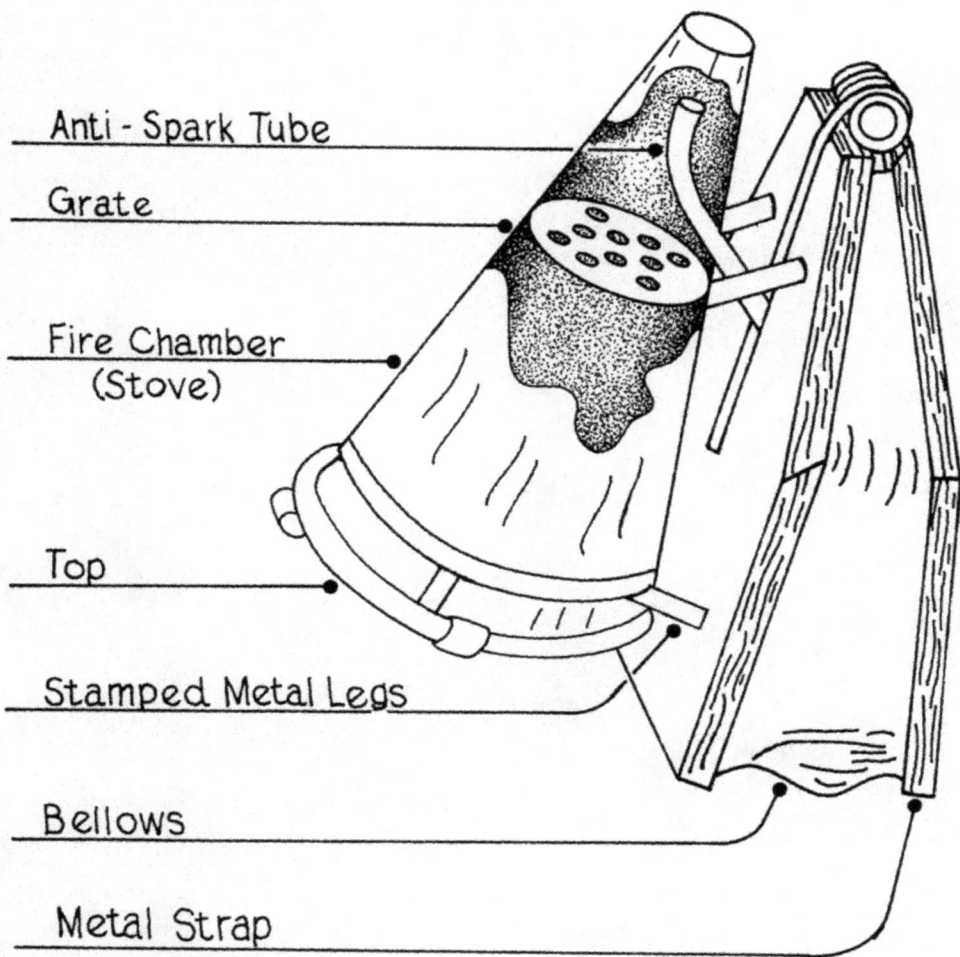

Anti - Spark Tube

Grate

Fire Chamber
(Stove)

Top

Stamped Metal Legs

Bellows

Metal Strap

www.ingramcontent.com/pod-product-compliance
Lightning Source LLC
Chambersburg PA
CBHW081157090426
42736CB00017B/3358